Contents *continued*

Contents *continued*

Contents *continued*

This booklet was prepared by an expert sub-committee of the Health, Safety and Environment Committee (HSEC) of the Royal Society of Chemistry. The members of the sub-committee were:

Mr K Everett

Mr R W Hazell (Secretary)

Mr S G Luxon (Chairman)

Dr R Owen

Mr D M Sanderson (Vice-chairman)

Mr H G E Wilson

Dr I Wrightson

This guide dealing with safe practices in chemical laboratories has been prepared by the Royal Society of Chemistry. It is essentially practical in its outlook and comprehensive in its scope. General principles of health and safety are well defined and the means whereby they can be translated into effective action are clearly described.

I have both pleasure and confidence in commending this guidance to all those who manage or who work in chemical laboratories. They can make an important contribution to the safety and health of all concerned by applying the advice contained in this guide.

A J LINEHAN
HM Chief Inspector of Factories

Section A – Introduction

This booklet was prepared by an expert sub-committee of the Health, Safety and Environment Committee of the Royal Society of Chemistry. It is the successor to the "Guide to Safe Practices in Chemical Laboratories" which itself derived from the "Code of Practice for Chemical Laboratories" published by the Royal Institute of Chemistry in 1976.

The broad objective of the current booklet remains the same as that of its predecessors: to provide general guidance on which specific procedures can be based and to point out relevant statutory requirements. Like its predecessors, it recognizes that the design of laboratory facilities, equipment and procedures is subject to change, particularly as a result of developments in legislation. The authors have therefore attempted to anticipate some such developments and allowance has been made for legislation known to be pending.

There can be considerable differences in the nature and scope of the work carried out in chemical laboratories. Furthermore there are obvious differences between large and small organizations in the scale of resources available for health and safety advice. Thus the booklet is concerned more with the principles of laboratory safety than with the provision of detailed advice on particular activities. The approach recommended is the establishment of "Local Rules", that is procedures based on the general principles presented here and adapted to the specific needs of the particular workplace. It is intended that this booklet should be read in conjunction with the Society's published guidance on the Control of Substances Hazardous to Health (COSHH) Regulations 1988 which contains more detailed recommendations (see Section F).

The necessity for those with managerial responsibility to maintain "current awareness" of legislation, codes of practice, etc cannot be over emphasised.

Although the most onerous duties rest with those possessing managerial accountability it must be stressed that responsibility for the health and safety of individuals lies with workers at all levels. A positive attitude to safe working practices is the responsibility of everybody involved.

Although directed primarily at members of the Society who manage or are employed in laboratories, the general principles outlined may be of more general interest insofar as they are applicable to staff in any laboratory where chemicals are used.

Where used in this booklet the following meanings should be assigned to the respective terms:

"Must" is used to indicate that the level of compliance required is equivalent to legislative requirements.

"Should" is used to indicate a level of compliance representing good working practice.

"Reasonable" where used indicates a proper practical balance between the costs and benefits involved.

The terms "acute" and "chronic" have their usual medical meanings.

Section B – Organization for Safety

A well organized and well managed laboratory is likely to be safe and pose minimal risks to health. The Health and Safety at Work Etc. Act 1974 (HSW Act) requires an employer to prepare and publish a written statement of policy for health and safety in any establishment with five or more employees. This must include details of the management structure and arrangements made for implementation of the policy.

THE SAFETY POLICY

(See also Section F)

The employer's safety policy is the key to high standards of health and safety in any laboratory. It is a statement of intent and should demonstrate management commitment at the highest level.

The policy must describe the organization which is responsible for its implementation. Although the employer is legally accountable for health and safety in the laboratory some functions may have to be delegated to others. Such individuals and their relationships form the organization, which should be clearly described. It should show a logical delegation of functions from management to local supervision and individual employees. Key employees should be identified in the statement and their duties clearly defined (job descriptions may be desirable in some cases). The relationships of any functional advisers (such as safety advisers, fire prevention officers, radiation protection advisers) with staff in positions of executive authority (eg managers of industrial laboratories or heads of department in educational institutions) should be defined.

Arrangements must be made to ensure that the employer's statement of intent is translated into safe working practices. These arrangements should include systems and procedures to cover the four main topics described in the HSW Act: the provision and maintenance of a safe place of work; the use, handling and storage of plant and substances; the provision of information, instruction, training and supervision; and consultation with the workforce. It is vital to achieve the co-operation of employees if the safety policy is to be effectively implemented.

In a laboratory the systems and procedures should take account of the following:

a) storage and use of hazardous materials and substances, (eg compressed gases, flammable liquids, toxic chemicals, radioactive chemicals)

b) maintenance and safe use of electrical equipment

c) safe handling of micro-organisms

d) safety of maintenance staff who might enter hazardous areas (eg by "permit to work" systems)

e) safety of employees who could be affected by maintenance work (eg work on safety systems)

f) arrangements for periodic environmental and other monitoring

g) emergency procedures in the event of fire, explosion, injury, spillage, etc.

h) arrangements to ensure the safety of contractors, visitors and others

i) arrangements for out-of-hours working

j) specific operator instructions for plant and machinery (eg use of guards)

k) specific safety training of staff

l) occupational health surveillance

m) provision of protective clothing

n) regular examinations of equipment such as pressure vessels and lifting gear.

Procedures should be developed to cover not only special hazards but also everyday hazards such as those associated with lifting and carrying. Many accidents occur to persons going about their routine tasks. Working procedures should be correctly documented and the documentation should be readily available at the workplace.

THE SAFETY ADVISER OR SAFETY OFFICER

The main function of the safety adviser should be to provide management and laboratory staff with professional advice and information on safety matters. In establishments with several laboratories or departments the co-ordinating role of the safety adviser becomes important. Other roles that the safety adviser could fulfil include:

a) organization of safety training

b) liaison with outside bodies concerned with safety (eg the Health and Safety Executive (HSE), Fire Authority,

c) provision of technical expertise and resources (eg access to environmental monitoring, hazard and operability studies)

d) investigation of accidents and dangerous occurrences

e) maintenance of records (eg accidents and dangerous occurrences statistics)

f) monitoring of safety inspections

g) liaison with other specialist advisers as appropriate

h) to make assessments of potential risks and the need for preventative action.

It must be stressed that safety advisers should never accept or be assigned managerial responsibility for health and safety.

CONSULTATION WITH EMPLOYEES

In every laboratory there will be individuals and small groups who have a special knowledge of the work being carried out. Their co-operation in and contribution to the establishment of safety arrangements and safe systems of work is essential.

A recommended way of involving employees is to set up a Safety Committee consisting of members representing all interests. Where employees are members of a recognised Trades Union, the Safety Representatives and Safety Committee Regulations 1977 provide a legal framework within which employers and Trades Unions can make arrangements for the functioning of the safety committee.

The main object of a Safety Committee should be the promotion of communication and co-operation in developing safety arrangements. Its functions could include:

a) review of the accident, dangerous occurrence and occupational ill-health experience of the laboratory or establishment

b) consideration of reports from the safety adviser, the enforcement inspectors, representatives of employees and other interested parties

c) agreement of draft safety rules, working procedures and systems of work.

The functions of Safety Representatives appointed under the Regulations include investigations of potential hazards and dangerous occurrences at the workplace, investigation of complaints by any employee and inspection of the workplace. In the absence of recognised Trades Unions, employees should be invited to provide adequate representation when matters of health and safety are considered.

INFORMATION, INSTRUCTION AND TRAINING

Arrangements for the provision of information, instruction and training for health and safety must be adequate and apppropriate for all staff bearing in mind the nature of their tasks. In addition to this general requirement special training is necessary under the COSHH Regulations (Reg. 12). Therefore the needs of all levels of staff should be identified and training programmes developed accordingly. Training should take account of the operation of local rules. Where possible health and safety training should be integrated with job training.

Employees with specific safety duties should receive adequate training to allow them to carry out these duties effectively.

STUDENTS AND OTHER TRAINEES

Section 3 of the HSW Act requires employers to ensure the health and safety of persons who are not their employees but who may be affected by their work activity. This includes students.

The level of knowledge and skill of the student or trainee will be less than that of a fully trained person and the degree of guidance and supervision required will therefore be higher. If the expectation is that students or trainees are likely to be more careless than others then reasonable steps must be taken to protect them from their own carelessness. On the other hand students have a duty to take reasonable care for staff and anyone else they can reasonably foresee as likely to be affected by their negligence.

SAFETY AUDITS AND INSPECTION PROCEDURES

In any organization it is essential to assess the effectiveness of the policy and arrangements for health and safety. There is no single indicator of performance and monitoring procedures should cover:

a) records of accidents and ill-health

b) the extent of compliance with legal requirements and codes of practice, etc. relating to health and safety

c) the extent of compliance with relevant systems and protocols.

Various techiques of audit and inspection have been developed. Both periodic and ad hoc spot inspections should be made.

Techniques for monitoring and assessing safety vary in complexity from simple visual observations through inspection and auditing to detailed examination of the hazards associated with premises, plant, material, operations and systems of work. Such examinations may result in a quantification of the risk of the activity. The main techniques for monitoring and assessing safety in the laboratory are outlined below.

The simplest monitoring technique is the safety tour. This is little more than a hazard spotting exercise and is designed to ensure that for example an acceptable standard of housekeeping is achieved, obvious hazards are removed and generally good standards of health and safety observed. Safety tours are usually carried out by management and may be unannounced.

A more detailed approach is to carry out safety inspections. These are usually carried out at regular pre-planned intervals by managers who may be accompanied by safety representatives. Inspections are designed to ensure that safe operational precedures and systems of work are followed in addition to ensuring that good standards of health and safety are observed. This is a useful technique. It may reveal major problem areas requiring attention especially if the reasons behind the failures are fully investigated.

A safety audit is a systematic critical examination of work activities in order to identify potential hazards and determine the associated levels of risk. The objective of an audit is to ensure that the organization's health, safety and loss prevention measures are operating effectively and that they are meeting legal requirements. Audits include the examination of the laboratory safety policy, organizational and operational procedures and the design, layout and construction of the laboratory and its equipment. Auditing often makes use of a detailed questionnaire to identify organizational strengths and weaknesses. Following an audit an action plan should be drawn up. Monitoring should then be carried out to ensure progress in those areas found to require attention.

There are a number of formalised safety auditing systems available such as those provided by the Royal Society for the Prevention of Accidents and the British Safety Council. In addition there are a number of health and safety consultants who operate their own auditing and monitoring schemes. Some large companies (especially from the chemical industry) also offer a commercial service.

The above monitoring techniques may reveal major uncertainties which need more detailed examination and assessment using procedures such as hazard and operability studies (HAZOP), hazard analysis (HAZAN) and quantified risk assessment (QRA).

A HAZOP study is a structured critical examination of plant or equipment in order to identify all possible deviations from an intended design along with the consequent undesirable effects for safety and operability. The possible deviations are generated by rigorous questioning prompted by a series of guidewords applied to the process conditions.

HAZOP studies are carried out by a small team of people chosen for their individual expertise in design, operations, maintenance or health and safety. It is a technique used for the elimination of the hazards which are identified rather than simply reducing the likelihood of their occurrence. Although HAZOP studies are used primarily for thoroughly checking a design they can be used for existing plant and equipment. In the laboratory HAZOPs can be successfully applied to the design of experimental or pilot plant equipment.

HAZAN is the identification of undesired events which can result in a hazard, the analysis of the mechanisms by which these events could occur and the estimation of the extent, magnitude and likelihood of any harmful effect. The analysis of the mechanisms leading to these events can be carried out using logic diagrams which can give qualitative information and provide a model for quantification. Logic diagram systems include fault-tree analysis, event-tree analysis and cause-consequence analysis. Both fault-tree and event-tree analyses can provide the basis for deriving the numeric probability of a certain event or risk. HAZAN techniques can also be applied to experimental or pilot plant equipment.

A further technique is quantified risk assessment (QRA). It is an evaluation of the likelihood of the undesired events and the consequence of harm or damage along with value judgements on the significance of the results. QRA is used extensively in the chemical and petrochemical industries particularly when large inventories of hazardous chemicals are stored and used.

EMERGENCY PROCEDURES

Firefighting

All employees must be trained to deal promptly with any fires that break out in the laboratory. It is essential that sufficient appropriate extinguishers are available and that employees are familiar with their use and limitations. Extinguishers are provided to put out small fires or to contain larger fires while occupants escape from a room or building to a place of safety. With this in mind employees should be trained to decide between the competing needs of tackling the fire and raising the alarm. Such conflict may not arise if more than one person is available (see below).

The choice of fire extinguisher depends on the nature and scale of the laboratory operations. For small laboratory fires which involve flammable liquids (Class B fires) the choice of portable extinguishers lies among vaporising liquids, carbon dioxide and dry powders. In most laboratories the first two are likely to be preferred. Any of them may be used on electrical fires. Irrespective of the choice of extinguisher everyone should be evacuated from the location at which a fire has occurred because the combustion products of even commonplace materials may be highly toxic.

Fires involving large quantities of contained flammable liquids are best extinguished with foam or dry powder. For fires involving spilled or running flammable liquids dry powder is the most suitable extinguisher. Foam should not be used on electrical fires.

For fires involving solid materials such as wood or paper (Class A fires) water is the preferred extinguishing medium because of its cooling power. Water should not be used on fires involving electrical equipment and care should be taken in laboratories where water-reactive chemicals are in use. Nevertheless, water extinguishers should always be available to complement other types.

Fires involving gases or vapours (Class C fires) can be very difficult to fight as extinguishing the flame can allow accumulation of a dangerous gas/air mixture which is liable to re-ignite or even explode should it reach an ignition source. The supply of gas should be isolated before the flame is extinguished with a suitable extinguishing medium such as dry powder or foam.

Particular risks are associated with fires involving water-reactive substances such as certain metals, metal hydrides or metal alkyls (Class D fires). Special powders are available for use in such situations.

Some laboratory operations are best protected by fixed fire extinquishing systems, which may be either activated automatically or operated manually. Extinguishing media used include water sprays, halons and carbon dioxide. The protection of employees must always be carefully considered whenever fixed automatic systems are installed, bearing in mind the potentially asphyxiating properties of such systems. Advice should be sought from the HSE and from the fire brigade with whom close liaison should be maintained.

People whose clothes are on fire

In such cases the first requirement is to extinguish the fire. It is especially important to keep the flames away from the head. Minor clothing fires can usually be dealt with by placing the victim horizontally and wrapping in a blanket of adequate size (minimum 0.9m × 0.9m; maximum 1.8m × 1.8m) provided for the purpose. However, clothing fires can induce panic and such an approach may be difficult. This is particularly so in clothing fires accompanied by spillage of burning liquid where the use of a blanket to extinguish the fire may not be feasible initially due to the fierceness of the fire. Portable fire extinguishers provided for the laboratory should be used so that the victim can be reached and the clothing fire dealt with.

Rescue

An appropriate number of laboratory staff should be trained in rescue procedures. An example of the kind of situation envisaged is where someone has been overcome by smoke during a laboratory fire, or by toxic gas or fumes as a result of an explosion, and has to be rescued from the laboratory area by a colleague. The main equipment needed is an approved self-contained breathing apparatus of adequate capacity. The person who wears it must be adequately trained and capable of effecting a quick rescue of a casualty. Such equipment can also be used to enter a toxic atmosphere created by a laboratory accident in order to shut down services and/or processes and to open windows. When a rescue is being effected the rescuer should take precautions to avoid becoming a casualty himself. This may entail having a second trained rescuer standing by where this is practicable to deal with any emergency that may arise. In such cases the second rescuer should have his own breathing apparatus.

Spillages

All employees should be familiar with appropriate procedures for dealing with any spillages of the substances handled in their laboratory – washing down, vacuuming, adding sand, sweeping up etc. Supplies of necessary absorbents or decontaminants should be readily available together with instructions for their use and appropriate protective equipment for the user.

Training procedures and evacuation

Clearly understood arrangements should be laid down for emergency evacuation of laboratories. Line managers should make the evacuation arrangements appropriate to each area and organize frequent evacuation exercises. All persons given special responsibility for evacuation procedures should have deputies who should be available to act in their absence. Fire alarms should have a distinctive sound and should be audible to all those at risk. Alarms should be frequently tested, preferably at least once a month.

Escape routes should be distinctly marked as such. They should not be obstructed, even temporarily. All personnel should be aware of the locations at which to assemble after evacuation. Suitable precedures should be established to ensure that all personnel, including visitors, are accounted for in an emergency.

RECORDING AND REPORTING ACCIDENTS AND INCIDENTS

Accident and incident reporting procedures are a valuable aid to the management of health and safety. Each establishment should have a system for the prompt reporting and recording of all accidents and incidents. The information so collected may be used:

a) to discover cause, to prevent recurrence and to determine whether there are any wider implications for safety

b) to identify hazards not otherwise foreseen

c) to compare statistics with those of other laboratories to identify trends.

All accidents and incidents no matter how trivial they may appear to be should be recorded. However it must be noted that there is a statutory requirement to report certain categories of incident and accident. Further details of these statutory requirements are given in Appendix I.

MEDICAL AND HYGIENE SERVICES
(see also Section E)

Medical services

It is essential for the laboratory manager to review operations in the laboratory in order to identify the risks and to establish appropriate procedures to maintain the health of employees. In order to do this advice may be needed from a medical practitioner trained in occupational health care. Laboratories forming part of a larger organization will probably have access to an in-house medical adviser but, in the UK, smaller laboratories can obtain advice direct from the Employment Medical Advisory Service (EMAS) of the HSE.

The scope of the medical services provided will depend on the nature and scale of the laboratory operations. Minimum standards of first aid services are defined in the Health and Safety (First Aid) Regulations 1981, which impose a duty on employers to provide "such equipment and facilities that are adequate and appropriate to the circumstances".

Further notes on first aid procedures are given in Section E.

First aid

The majority of injuries arising from laboratory accidents require rapid first-aid treatment, eg splashes of corrosive chemicals into the eyes or on to the skin, burns, cuts, inhalation or ingestion of corrosive or toxic materials, electrocution. Thus in the event of a laboratory accident the immediate application of appropriate first-aid treatment can be vital. The Health and Safety (First Aid) Regulations 1981 provide a legal framework within which each establishment or laboratory can develop effective first-aid arrangements suitable for its own needs. The Regulations are supplemented by an approved Code of Practice and Guidance Notes which give general advice on the recommended levels of provision of first-aid personnel, equipment and facilities.

Every employee must have reasonably rapid access to first-aid and emergency services. The equipment and facilities provided will vary according to circumstances from small first-aid kits to a first-aid room. The needs of persons such as cleaners, maintenance staff and laboratory staff working late or in off-site locations should be considered.

First-aid boxes should be designed to be used not only by trained first-aid personnel but also by any employee who might suffer injury. They should therefore contain only material which may be used by unqualified persons without further risk of injury. The HSE has prepared a list of recommended contents. If additional equipment such as special antidotes or resuscitation apparatus is necessary to deal with any special hazard it should be kept near to the point of potential use and used only by trained persons.

In all but the very smallest establishments one or more trained first-aiders should be available.

First-aiders should have attended a course of general first aid training approved by the HSE, together with additional training on those particular aspects of first aid necessary to deal with the special or unusual hazards existing in the laboratory.

In addition all laboratory personnel should be familiar with the elementary principles of laboratory first aid, particularly the treatment of splashes of corrosive liquids into the eyes.

Occupational hygiene and medical surveillance

Occupational health care depends on identification of potential hazards to the health of workers and the control of exposure to such hazards to within acceptable limits. Thus it is the responsibility of the manager, irrespective of the size of the laboratory, to arrange for an assessment to be made so as to:
- identify the hazards
- quantify the risks to health
- determine appropriate control measures.

The manager must subsequently arrange for:
- the provision, use and maintenance of the necessary control measures
- monitoring of the workplace as may be necessary to ensure compliance with agreed standards (this may include biological monitoring)
- monitoring of the health of the employees as appropriate
- a regular review of all assessments and procedures.

The hierarchy of measures for controlling exposure is set out in paragraph 34 of the COSHH Approved Code of Practice (see Section F).

Health surveillance will in general only be required if "the exposure of the employee to a substance hazardous to health is such that an identifiable disease or adverse health effect may be related to the exposure, there is a reasonable likelihood that the disease or effect may occur under the particular conditions of his work and there are valid techniques for detecting indications of the disease or the effect" (COSHH regulation 11(2)(b) but see also COSHH 11(2)(a) and COSHH Schedule 5; see Section F).

Where there is a significant risk to health prudent managers should seek professional occupational health advice in respect of activities for which they are responsible.

In large organisations such advice may be available in-house. In other cases it may be advisable to seek help from an independent consultant. The Employment Medical Advisory Service (EMAS) of the HSE provides advice on medical surveillance.

Whatever forms of environmental or personal monitoring and medical surveillance are adopted it is essential that accurate records are kept (see Section B).

DOCUMENTATION FOR EPIDEMIOLOGY

Epidemiology deals with the incidence of disease in groups of individuals. It can yield valuable information on the relationship between the work environment and the health, morbidity and mortality of the people employed there. Epidemiological studies can identify occupational health risks. The technique employed is often retrospective and depends upon the availability of data accumulated over several years. The necessary data relate to the occurrence of disease, the nature of the occupation, and the level of exposure to toxic substances.

The laboratory manager should establish a system to record information about occupational exposure to hazardous substances. In essence this means that there should be a general record of who does what, where, when and for how long under what level of contact with which substances. It may be that records established for other purposes will furnish much of the information needed (eg dates of employment and of retirement for each individual).

Where new work entails the use of novel chemicals or known toxic materials it is advisable to consult specialists on the desirability and feasibility of biological monitoring of the workers involved. Information on the health of an individual must of course be confidential.

Epidemiological studies are undoubtedly costly in terms of time and resources but provide invaluable evidence and information on causes of occupational ill-health. Whatever studies are undertaken the co-operation of all personnel is essential.

SECURITY AND ACCESS CONTROL

Organizational arrangements designed to ensure a safe working environment require that certain security operations such as access control are co-ordinated with safety administration. Within the security field safety aspects can be divided into four distinct areas:

1. Employers have a duty to ensure the safety of all visitors (including contractors). Thus visitors need to be informed of hazards to which they may be exposed and of appropriate safe-working procedures. It is therefore essential to control access to premises. Such control is normally achieved by ensuring that visitors report to specific locations where information and any necessary safety equipment (eg eye protection) may be provided. Visitors should not normally be permitted to move unaccompanied about the site. (It is important to remember that "visitors" include maintenance engineers and salesmen.)

2. Reasonable steps may need to be taken to protect employees from acts of vandalism or violence. The degree of hazard obviously varies with the type of operation but certain operations are nowadays at a high risk. An appropriate policy should be established in conjunction with security staff where they exist. If necessary advice should be sought from the police.

3. The posibility of arson is not always adequately considered in the design of fire precautions.

4. Although strictly an access control problem, the question of access by employees to hazardous substances or equipment is considered separately in the following section.

Good practice dictates that all chemical and physical hazards should be identified and brought to the attention of all personnel so that they can be involved in the promotion of a safe and healthy working environment. Hazard data should be obtained for all agents used and steps should be taken to eliminate the risks to personnel as far as practicable. Strict procedures should be drawn up to ensure that any residual risk is minimised. All work should be supervised to ensure that the necessary precautions are followed. Safety audits and inspections should be carried out as necessary to ensure that hazards are identified and risks properly assessed so that adequate precautions can be implemented.

CHEMICAL HAZARDS

General

Although all substances are hazardous to health and/or safety under certain conditions the materials discussed in this Section present risks over a wide range of normal conditions. They usually have one or more of the following characteristics –

low flash point

spontaneous reactivity

powerful oxidizing or reducing properties

severe irritant or corrosive properties

high acute or chronic toxicity

carcinogenic or teratogenic properties.

In this Section hazardous materials are grouped under the broad headings of flammable, explosive, corrosive and irritant, and toxic substances. Radioactive materials are the subject of special legislation and are not dealt with in this booklet.

Flammable substances

The is a high fire risk in many laboratories because of the storage and handling of flammable substances, which may be in solid, liquid or gaseous form. Particular attention should be paid to fire prevention in both design and operation of laboratories (see Sections D and E). A significant general hazard that is often disregarded is a tendency to obstruct the means of escape from the workplace. It is essential that steps are taken to ensure that this does not occur.

The magnitude of the hazard from flammable substances depends both on their physico-chemical properties and the quantities involved. For a liquid the flash point, auto-ignition temperature, explosive limits, vapour density and ability to accumulate an electrostatic charge are all important factors. The possibility of such hazards being accentuated through oxygen enrichment should also be considered.

Almost all flammable vapours can be ignited by common ignition sources such as flames or incandescent surfaces but some substances (eg certain ethers, aldehydes and carbon disulphide) can ge ignited at lower temperatures such as those produced by hot plates, ovens and isomantles. Carbon disulphide is particularly dangerous in this respect because its auto-ignition temperature is only just above 100°C.

Fire or explosion can occur when flammable vapours are mixed with oxygen in proportions within critical values known as the Lower and Upper Explosive limits (LEL and UEL, respectively). For most solvents the LEL lies in the range of 1-5% in air and therefore good ventilation is essential in order to eliminate the risk of forming a flammable or explosive atmosphere when such substances are used. However, it is significant that the LEL is considerably greater than the recommended hygiene standards for the concentration of vapour in the workroom air.

Vapours from flammable liquids are denser than air and thus tend to sink to ground level where they can spread over a large area. Care should be taken to minimise the production of such vapours and the associated risk of ignition by flashback from a remote source. Flammable vapours may also be ignited by sparks caused by the discharge of static electricity. (Static charges may be generated in liquids flowing through pipes or by their agitation or stirring). When large quantities of flammable solvents are dispensed it is good practice to ensure that all vessels used are of metal and are electrically connected and earthed.

Quantities of flammable substances in laboratories should be kept to a minimum. When not in use solvents are best kept in suitable fire-resistant cabinets. Larger quantities of solvents should be stored in correctly designed ventilated stores. Certain substances having flash points of 21°C or below (including many common solvents) come within the scope of the Petroleum (Consolidation) Act 1928, and stores of such materials are subject to licensing by the Local Authority. In premises covered by the Factories Act 1961, regulations limit to 50l the total quantity of highly flammable liquid which may be kept and the maximum size of any container on the bench to 500 ml (Highly Flammable Liquids and Liquefied Petroleum Gases Regulations 1972). These and other relevant items of legislation are listed in Section F.

When flammable liquids are transported within the laboratory appropriate carriers should be used for other than small glass bottles. The dispensing of flammable liquids from bulk containers should be carried out by trained staff in a safe area (see Section D). All containers must be labelled with the names of the contents and appropriate statutory markings. Flammable solvents should not be poured down sinks.

Flammable solids often encountered in laboratories include the alkali metals, magnesium metal, metallic hydrides, some organometallic compounds, phosphorus and sulphur. Such substances must be stored safely in correctly labelled containers and staff must be made aware of the hazards associated with their use.

Explosive substances and mixtures

The risk of explosions from mixtures of flammable vapours with air has been referred to above. Explosions can also occur when flammable substances are mixed with solid or liquid oxidants and the need for such combinations should always be assessed before proceeding.

Some substances can detonate as a result of friction, shock, heat (eg organic peroxides) or contamination (eg mixtures of perchloric acid with a wide variety of materials can be highly unstable). Attention must be given to the storage instructions which must be provided with such materials. In addition to hazards associated with the handling of these materials it should be remembered that unstable substances may be formed during chemical operations or prolonged storage. Appropriate measures should be taken to avoid this occurring. For example, certain ethers, alcohols and aldehydes can form peroxides which may detonate during distillation. For this reason bottles of materials prone to peroxidation should not be kept for prolonged periods once they have been opened.

Some substances are inherently unstable and can detonate under certain conditions of pressure and temperature (eg acetylene, carbon disulphide and substances prone to autopolymerisation).

Corrosive and irritant substances

Many chemicals in common laboratory use are corrosive or irritating to body tissue. They present a hazard to the eyes and skin by direct contact, to the respiratory tract by inhalation or to the gastro-intestinal system by ingestion.

Corrosive chemical liquids represent a very significant hazard because skin or eye contact can readily occur from splashes. The intact skin is an effective barrier to many irritant chemicals and splashes thereon cause only minor irritation if treated promptly. However, the eyes are very vulnerable. It is therefore essential that approved eye protection be worn in all laboratories where corrosive or irritant chemicals are handled (see Section E).

Corrosive and irritant gases and vapours are hazardous to all parts of the body; certain organs (eg the eyes and the respiratory tract) are particularly sensitive. The magnitude of the effect is related to the solubility of the materials in the body fluids. Highly soluble gases (eg ammonia, hydrogen chloride) cause severe nose and throat irritation, whilst substances of lower solubility (eg nitrogen dioxide, phosgene, sulphur dioxide) can penetrate deep into the lungs. Nose or throat irritation may be absent with some substances and therefore cannot be relied upon as a warning.

Skin disease caused by occupational exposure to irritant chemicals is commonly either contact or sensitisation dermatitis. The former refers to an inflammation of the skin caused by direct contact with the irritant substance. Liquids and solids usually affect the hands and forearms whereas gases, vapours, fumes and dusts can affect all exposed skin surfaces including the face and neck. However it is important to remember that transfer of irritant materials (eg from contaminated gloves) can affect other parts of the body.

Sensitisation dermatitis results from an immune response to the irritant substance and symptoms seldom appear immediately after exposure. Unlike contact dermatitis the area of skin affected is usually extensive and occasionally the whole body surface may be involved. Many substances have been associated with sensitisation dermatitis including certain amines, nitro-compounds, nickel compounds and materials such as epoxy and phenol-formaldehyde resins. Unfortunately owing to variations in individual susceptibilities it is not easy to predict sensitisation and it is always prudent to minimise contact with all chemicals whether or not they are suspected of causing sensitisation. Safe handling procedures are considered in Section E as are first aid measures and procedures for dealing with spillages. In all cases of body contact with corrosive chemicals speed of response is vital if permanent injury is not to result. Employees must be trained to react quickly and correctly if an incident occurs (see Section E).

Toxic substances

The nature and magnitude of toxic effects will depend on many factors including the nature of the substance, the route of exposure, the magnitude of the exposure, the frequency and duration of exposure, and individual susceptibility. All chemicals are potentially hazardous to health. The degree of hazard can be estimated from a knowledge of the toxicity, the likely exposure and the results of previous experience if these are available. Toxicity data may be gained from relevant research or development, from the results of laboratory testing on animals or other organisms, from epidemiological studies of the incidence of disease or causes of death in certain population groups or by analogy with the properties of other chemicals of similar structure. However, such data should be used with discretion. It should be noted that warning labels on containers may highlight only the most serious hazards of the substance they contain. Substances may present hazards other than those indicated.

The fact that a substance is known to have toxic properties does not imply that it cannot be used. On the contrary such knowledge is essential for the specification of appropriate precautions to ensure a safe and healthy place of work. Even substances of unknown toxicity may be handled safely if suitable precautions are taken. Such precautions will depend on the nature of the substance and the manner of its intended use, and are thus specific to each individual application.

Toxic effects may be acute or chronic, local or systemic. Acute effects usually result from short-term exposure and are often reversible; chronic effects are usually the result of long-term or repeated exposure and are sometimes irreversible. Local injuries are confined to the immediate area of exposed tissue; systemic damage can occur in any organ after the toxic material has entered the body.

Inhalation of gases, fumes or vapours and subsequent absorption through the lungs is a rapid route into the bloodstream. The rate of absorption depends on factors such as the respiration rate, the atmospheric concentration of the toxic substance and its solubility in tissue fluids.

Dusts and fibres can also be inhaled; their fate depends on their size. Particles smaller than about 0.1 µm behave as vapours; those larger than 5 to 10 µm are generally caught in the upper respiratory tract whence they are usually carried up and swallowed. Particles of intermediate size present the greatest hazard as they remain deep in the lungs.

Some chemicals may be absorbed through the intact skin and can produce serious systemic effects. (See also "Corrosive and irritant substances").

Common causes of accidental ingestion of toxic substances in the laboratory or workshop include smoking (handling of cigarettes with contaminated hands), use of cosmetics, eating or drinking, biting of fingernails, licking of labels and oral pipetting. Ingestion of inhaled dust particles can also occur (see above).

Depending on the mode of entry, the nature of the chemical and the magnitude and frequency of the dose, toxic substances may be excreted in varying amounts from the body (for example in the urine or faeces, or by perspiration or exhalation). Whilst in the body chemicals may be metabolised into less harmful substances (detoxification) but in some cases they can be converted to more toxic products (eg ethylene glycol is oxidized to oxalic acid).

The effects of toxic substances range from minor irritation of skin or eyes to serious irreversible changes and death. Much attention has been given to the carcinogenic potential of certain chemicals but mutagenic, teratogenic or allergenic properties must not be overlooked.

Suppliers of chemicals are now obliged by law to provide adequate information to enable their products to be handled safely. However such information may not always be complete and it may be necessary to adopt handling procedures that assume the highest levels of toxicity.

A strategy for the contol of toxic substances is provided by European Communities legislation (Council Directive of 27 November 1980 on the Protection of Workers from the Risks Related to Exposure to Chemical, Physical and Biological Agents at Work (80/1107/EEC) which forms the basis of COSHH. (See Section F).

Safe working with toxic substances depends on reducing the exposure at least to a level that produces no adverse effects. For particularly toxic substances the preferred method is to simply avoid their use by substituting less toxic chemicals. Unfortunately this may not always be feasible. Nonetheless operational control procedures must always be designed to reduce employee exposure to the lowest levels that are reasonably practicable; if such levels are still judged to be too high it may be necessary to abandon the work. The control procedures employed usually involve the use of containment facilities (eg fume cupboards or glove boxes – see Section D). As a last resort it may be necessary to wear approved personal protective equipment (see Section E).

In a number of hygiene standards attempts have been made to define levels of inhalation exposure that are safe and without risk to health for most workers. Hygiene standards in no way define an absolute limit between safe and unsafe conditions, therefore known individual susceptibilities should also be considered. Hygiene standards for certain substances are listed in Guidance Note EH40 from the HSE (see Section F). However, "the absence of a substance from the lists . . . does not indicate that it is safe. In these cases, exposure should be controlled to a level to which nearly all the population could be exposed, day after day, without adverse effects on health. . . . In some cases there may be sufficient information to set a self-imposed working standard, eg from manufacturers and suppliers of the substance, from publications of industry associations, occupational medicine and hygiene journals." (COSHH Approved Code of Practice paragraph 29. See Section F).

Whatever standards are accepted it should be remembered that toxicology is a developing science and that as new information becomes available criteria may change. As indicated above all procedures should be designed to minimise contact with chemicals irrespective of their known toxic effects. Prudent managers will review at regular intervals the handling of chemicals in their laboratories and establish appropriate programmes of occupational hygiene and medical surveillance for their staff (Section E).

PHYSICAL HAZARDS

General

In a chemical laboratory hazards may arise from various physical factors. These include noise, vibration, radiations and electricity. These factors are not covered in detail in this booklet. However they should always be considered and appropriate advice sought as necessary.

Certain other physical hazards are particularly associated with chemical laboratories. The most significant of these are dealt with below.

Extremes of temperature

a. Thermal Stress

Thermal stress can arise from working at high or low temperatures, particularly in enclosed environments. In such circumstances appropriate expert advice should be sought.

b. Cryogenic Liquids

The hazards associated with cryogenic liquids are mainly due to their extremely low temperatures but their flammability, toxicity or corrosivity should never be overlooked.

Spillage of cryogenic liquid on to unprotected skin can produce severe burns. Burns can also result from touching equipment cooled to low temperatures. Appropriate protective clothing should always be worn when handling liquefied gases if there is any danger of splashing (see Section E). Prolonged inhalation of cold vapour or gas should be avoided.

The possibility of oxygen deficiency should be considered if cryogenic liquids are stored or used in a confined or poorly ventilated space. In particular, they should never be transported accompanied by persons in a passenger lift because of the risk of asphyxia should lift failure occur.

One particular hazard associated with liquid nitrogen and liquid helium should also be considered. Both have a boiling point below that of liquid oxygen and there is a possibility that oxygen may condense from the atmosphere and cause either oxygen deficiency, or oxygen enrichment on subsequent boil off.

Liquid air and liquid oxygen can be hazardous if they come into contact with organic or other oxidizable materials; liquid oxygen is particularly dangerous in this respect. Explosions have occurred when wood soaked in liquid oxygen has been mechanically shocked.

All persons required to handle cryogenic liquids should be either experienced in their use or under the direct supervision of an experienced person. Training is needed in such operations as filling containers, transferring materials, venting storage vessels and dealing with condensates.

Non-ionising radiations

(i) Lasers (including lasing diodes)

Lasers should be used only in accord with an accepted Code of Practice, several of which are generally applicable within the UK (see Section F).

Two points which merit special attention are perhaps not emphasised sufficiently in the available Codes. First under normal operation the beam should be contained and personal eye protection should not be relied upon. Nevertheless there are certain circumstances (eg beam alignment) in which it is necessary to inspect the beam, and appropriate personal eye protection must then be worn.

Second, there is often a considerable hazard from the power supply to the laser. Therefore, the electrical portion of such equipment must be protected by interlocks. Repair and modification of the electrical supply should be undertaken by a competent person fully equipped for such operations.

(ii) Other forms of non-ionising radiation (eg ultraviolet, infra-red, microwaves)

The use of equipment emitting any other forms of non-ionising radiation should be permitted only where the exposure of personnel in respect of both wavelength and intensity can be kept within safe limits. Unless limited by design (eg by total enclosure systems) the escape of radiation should be monitored on a regular basis using equipment appropriate for both intensity and wavelength measurements.

HAZARDOUS TECHNIQUES

There are some operations in chemical laboratories that can give rise to other hazards. The most significant of these are considered below.

Oxidation experiments and other reactions involving explosion risks

While this Section refers primarily to oxidation experiments the advice given is equally applicable to other types of reaction that carry explosion hazards.

The scale of all oxidation work, unless known to be safe, should be kept to a minimum. Staff should know the flammable limits (sometimes incorrectly referred to as "inflammable limits") of gaseous mixtures and, if possible, should aim to work well outside the flammable or explosive range. It should be remembered that changes of temperature or pressure modify these ranges.

The magnitude of the maximum possible energy release and the rate of release should always be considered. Where the apparatus is not capable of containing such release it should be placed within an adequate enclosure from which personnel are excluded throughout the period of risk. Safe operating procedures should be considered including the siting of controls outside the enclosure and if appropriate the provision of interlocks.

When flammable gases are taken outside the enclosure (eg for analysis or to exhaust) they must be made safe, for example, by dilution with an inert gas to below the Lower Explosive Limit. Mixing of gases should be carried out inside the enclosure. Appropriate pressure relief vents and flame traps should be incorporated in the apparatus and adequate maintenance procedures must be adopted.

Indicating liquids used in apparatus should be inert to reactants. For example, organic indicating liquids (eg dibutyl phthalate) can be peroxidized with disastrous results. If reactive gases are to be mixed with inert gases a correctly designed mixing vessel should be used. Such a vessel will buffer line-pressure changes and variations in concentration. Exit gases should be rendered non-flammable, for example, by dilution with an inert gas.

Transfer of solids into vessels should ensure that traces of the solid do not remain on the vessel walls. Vapours may be ignited by pyrophoric materials (eg Raney nickel, platinum, cobalt, manganese, palladium). Many materials not normally pyrophoric may become so when present in a finely divided state.

Unstable peroxides may be formed as by-products or intermediates in liquid-phase oxidations, eg when certain ethers or aldehydes are oxidized. The person in charge of such experiments is advised where possible to determine an acceptable upper limit for peroxide concentration. The concentration of peroxide should be monitored to ensure that it remains below the limit.

The person in charge of the work should see that regular checks are made to ensure that the nature and degree of any explosion hazard do not exceed those for which the protective equipment used was designed.

Reactions at elevated pressure

Pressure vessels and their associated equipment must be designed and constructed to meet existing standards, and they must comply with any relevant statutory requirements. It is good practice and a normal insurance requirement to have a system of registration and labelling for pressure vessels and relief devices.

Inspection and/or testing should be carried out at appropriate intervals and detailed records kept. The design, purchase, construction and modification of such equipment should be undertaken only in consultation with a qualified engineer.

High pressure autoclaves should always be isolated in suitably constructed buildings. All pressure vessels should be equipped with a pressure guage, and a safety valve or bursting disc. The provision of adequate containment and venting facilities should be considered. (For example, the provision of a dump tank that will hold all the contents of the pressure system. Such a tank should be safely vented to atmosphere and placed in a safe area with the safety valve or bursting disc vented into it.) It is important that reactants and products should not attack any part of the pressure system and that they should be unable to discharge or escape into the workplace.

Where reactions involving flammable gases such as hydrogen are being carried out staff should ensure that there has been no unwanted accumulation of gas before lighting up or switching on electrical equipment. This requirement can be met either by testing the atmosphere or better by ensuring that the ventilation is sufficient to prevent such accumulation.

The maximum working temperatures and pressures for which a system was designed must never be exceeded. The equipment should be protected against any uncontrolled temperature rise. Operators should be protected by suitable shielding and remote controls should be provided. Operators should be alert to any difficulties in maintaining pressure. Such difficulties together with any evidence of corrosion or weak jointing should be reported and the equipment taken out of service.

Most sterilisers are pressure vessels and should be subject to the precautions outlined above.

Detailed written operational instructions and a line diagram should be provided in the control area of all vessels together with clear indications of safe working pressures. A work log giving operational history, including defects, maintenance and alterations, should also be provided. The attention of staff should be drawn to appropriate Codes of Practice (see Section F).

Glass handling

Many laboratory accidents arise from the handling of glassware. Staff should be thoroughly trained in the cutting, bending and annealing of glass tubing and in the assembly and dismantling of glass apparatus if such operations form part of their work.

The heating of glass flasks may also give rise to hazards. The safest and most effective way of heating glass flasks above the temperature of a water bath is to use an electric heating mantle. A heating mantle should be of a size appropriate to the flask being heated in order to ensure that the liquid level in the flask does not fall below the top of the mantle with resultant local overheating. Heating on a sand-bath over an open flame is unacceptable for flammable liquids. Oil baths may be used but should always be fitted with a thermostat unless under direct control. For teaching purposes the Bunsen burner under a wire gauze on a tripod is acceptable for non-flammable materials but the importance of slow initial heating should be stressed. Slow initial heating is also advisable in the use of a combustion tube in a gas train.

Detergents should be employed as the normal means of cleaning glassware. More drastic cleaning techniques such as the use of chromic acid (which is highly corrosive and toxic) should be used only when cleaning with detergents or solvents is inadequate. Such techniques should be covered by safe operating procedures.

Many injuries are caused by broken glass and sharp objects. Such injuries are not only dangerous in themselves. They may also provide a ready means for toxic substances and biological materials to enter the body. Injuries should be treated immediately. Safe working practices should be followed to minimize such hazards. Suitable containers should be provided for broken glassware and sharp objects.

Section D – Design

This Section should be considered in conjunction with Section E.

GENERAL PRINCIPLES

The design of laboratories is a compromise between what is desirable and what is economically or otherwise feasible. Nevertheless the following general principles should be considered by every manager whether he or she is planning a new laboratory or operating an existing facility.

Certain aspects of laboratory design are covered by legislation (see Section F).

All laboratories must provide appropriate facilities to enable employees to work safely and without risk to health. Laboratories should be well lit and provide adequate working space; the standards adopted will depend on the nature and scale of the intended activities. Due attention should be paid to ventilation.

Where possible the more hazardous substances and operations should be separated from those that are less hazardous. Small quantities only of those substances required for immediate use may be retained in working areas. Sites for the permanent storage of chemicals (including solvents and gas cylinders) should be separate from working areas.

In addition to washing facilities within the laboratory, provision should be made close to but outside the laboratory to enable employees to wash, eat, drink and smoke without additional risk of contamination from toxic chemicals or microbiological agents (see Section E).

Walls and floors should have smooth, impervious finishes, free from cracks that can harbour dirt and spilled chemicals. In some circumstances it is essential for all walls, floors and ceilings to have such finishes, eg in biochemical or radiochemical laboratories. In multi-storey buildings floors are best constructed of concrete, as this reduces the risk of accidental flooding from a room above. In laboratories where highly toxic substances are to be used it may be desirable to have non-openable windows to minimise interference with ventilation. If possible windows should be installed flush with the inner wall surfaces to minimise ledges which may harbour dust.

Benches should have impervious surfaces with appropriate resistance to chemicals and heat. Some traditional designs of laboratory benches are far from ideal, eg central shelving for stored reagents harbours dust and encourages reaching over other apparatus. Flat-topped benches with rear-mounted service outlets are preferable especially where controls are fitted at the front immediately below the bench-top. Sufficient service outlets should be available to minimise the use of trailing electrical cables and flexible piping. Trays of sufficient size to contain spillages should be provided for bench-top storage of those reagents needed for immediate operations. Other reagents should be stored in cupboards, which are always to be preferred to open shelves. Ventilation requirements for such cupboards should be considered.

There is no legally defined minimum free floor space requirement for laboratories. However, adequate space must be provided to allow safe and effective working. The following minima have been recommended:
- a total of 4.5 square metres per person
- 2.4 metres between island installations.

However both of these may have to be increased to take account of local circumstances. Adequate lighting should be provided to ensure that work can be carried out safely. It is important that lighting does not cause glare, adverse colour effects or too many dark shadows. Lights should not flash, flicker or cause stroboscopic effects. Good lighting has an important role to play in accident prevention. It assists in the recognition of hazards and can reduce the likelihood of visual fatigue and discomfort.

FIRE PREVENTION

Fire is a significant hazard in any chemical laboratory. Section C refers to the hazards from flammable substances, whereas the paragraphs below describe possible sources of ignition.

Fire depends on the presence of a fuel, an oxidant (usually air) and normally a source of ignition. Therefore, the prevention of fire relies on eliminating one or more of these. This can be achieved best by a combination of good design and safe working practice.

Two aspects of design are important. One demands that stocks of flammable substances are kept to a minimum, especially those outside approved storage areas. It will often be necessary for solvent stores to be licensed by the local authority (storage is considered in more detail below). The other requires provision for the containment of a fire in order to limit its spread to adjoining areas. Difficulties arise in laboratory buildings because of potential chimney effects created by fume cupboard exhausts and service ducts. Fire breaks are desirable in such ducts and fusible-link dampers may be necessary.

Ideally at least two protected escape routes should be provided. Emergency exits must be identified clearly and must not be obstructed by furniture or other moveable items. (Fire certificates are required for certain classes of premises).

An appropriate fire detection and alarm system should be provided. Smoke and/ or rate-of-rise heat detectors should be considered, especially in laboratories where equipment is left running unattended. Automatic fire extinguishing systems may also be considered. (The selection and use of portable fire extinguishers are discussed in Section E).

The design of safe working procedures is as important in fire prevention as the provision of adequate facilities (Section E).

STORES OF CHEMICALS AND GASES

Chemicals must be stored safely, preferably in a convenient, correctly designed store (see also Section D). Stocks should be controlled by careful management of purchase, stock rotation, use and disposal. Access to chemical stores must be restricted to authorised persons who have been instructed in the nature of the hazards and in appropriate safe handling precautions, including procedures in the event of spillage.

Containers must be labelled in accordance with legislative requirements (the Classification, Packaging and Labelling of Dangerous Substances Regulations 1984 –see Section F). In addition to identifying the contents the label may have to carry a hazard warning. Labels should display the sources of the material and the date of packing. This is particularly important for chemicals such as ethers that can form unstable products during storage.

The arrangement of chemicals in stores should be simple and logical (eg alphabetical). However, care should be taken to separate incompatible chemicals. In an alphabetic system this may be achieved by removing the chemical from the shelf and substituting a card or a marker which indicates where the chemical may be found. The card should also give brief details of the incompatibility.

Many substances are hazardous in their own right and special consideration should be given to their storage. Some are subject to legislative control, for example flammable liquids, medicines, controlled drugs and some carcinogens. Moreover many chemicals combine several hazards and careful thought should be given to their segregation. The appropriate fire fighting procedures should be carefully considered and clearly documented.

The principal hazards associated with the storage of flammable liquids are fire and explosion. These may arise, for example, from ignition of the liquid or from exposure of the storage installation to heat from a fire in the vicinity. Such hazards can be minimised by storing flammable liquids in properly designed containers inside fire-resisting store rooms or cupboards and by taking care to prevent or contain spillages during storage and dispensing. Sources of ignition should be removed from the store and no other combustible materials should be kept in it. Adequate provision of absorbent materials, eg dry sand, must be available to deal with spillages; an appropriate method for disposing of the used absorber must be arranged. On no account should flammable material be washed down drains if it is spilled.

Concentrated acids and alkalis should be stored at low level. Drip trays should be used and containers stored at floor level should be protected by a kick rail.

Suitable provision should be made for storage of waste chemicals before disposal (see also Section E).

Empty containers which have held flammable materials should be thoroughly cleaned and freed of all residual material before disposal to ensure that dangerous concentrations of vapour are not present in the vessel.

Chemical store rooms should be adequately ventilated at both high and low levels to prevent the accumulation of offensive, toxic or flammable vapours. Where substances are transferred or dispensed a separate area should be set aside with adequate facilities; including extraction and containment for spillages. Floors and shelves should be easily cleanable and resistant to chemical attack. Shelving should be rigid and stable. Attention is drawn to the requirements of the Petroleum Consolidation Act (see Section F).

Gas cylinders should be stored outside the laboratory. A secure open-air compound is an acceptable alternative to a correctly designed ventilated store room. Gas cylinders should preferably be stored vertically and secured to prevent them from falling. Acetylene cylinders must always be stored and kept upright. Full and empty cylinders should be segregated and individual gases kept in separate bays where possible. Flammable gases should be segregated from oxygen and other oxidizing gases. Where a gas distribution system is to be installed manifolds and pipes should be colour coded to accepted standards and labelled at appropriate points. Manifolds and pipes should be of appropriate material.

UTILITIES

General

It is essential that adequate services are available for laboratories. Due attention should be paid to their suitability, particularly where the nature or scale of the laboratory work involves changes from that for which the building was designed.

Electricity

All electrical equipment that may be exposed to the weather, to water, or corrosive, flammable or explosive atmospheres, or used in any special process, must be so constructed or protected to prevent danger arising from its use under such conditions. As some or all of these conditions may be found in laboratories it is essential that electrical services are designed and maintained to acceptable standards by competent persons.

Fuel gas

Almost all gas used as burner fuel in laboratories in the United Kingdom is natural gas (methane) from mains supplies. However, propane or butane from cylinders is sometimes used. All fuel gas lines (like other service lines) should be colour-coded to an accepted standard and labelled at intervals. The main gas supply to each laboratory building should be fitted with a readily accessible control valve to enable authorised persons to isolate the building.

Water

In addition to its use for washing, water is the principal heat transfer fluid for cooling or heating, condensers and water baths being familiar examples of its use. Water is also an essential first aid medium and should be readily available in all laboratory rooms, eg at sinks, eye washing fountains and emergency showers. Care should be taken to ensure the integrity of water supplies.

Non-potable water to laboratories should be supplied from intermediate storage tanks to minimise risks should the main supply be interrupted. Where drinking water is provided it must be adequately identified as such.

Steam

Steam may be found in development or pilot scale laboratories as a means of vessel heating. Steam lines should be lagged with suitable materials to minimise heat losses and the risk of both skin burns and ignition of certain highly flammable liquids. Any lagging exposed to mechanical damage should be protected.

Drainage

All drainage systems should be designed and constructed to accommodate the anticipated aqueous waste products of the laboratory (see Section F). The specification of the system, including details of the materials used, should be retained for reference purposes.

Compressed air

Compressed air may be supplied from cylinders outside the laboratory or from a compressor. When designing the system the air quality required should be considered as it may be necessary to provide facilities for drying or for the removal of oil mist. Appropriate pressure fittings and connections should be provided.

Where air supplies are required for breathing purposes a separate system should be provided which will ensure an adequate supply of sufficient purity at all times. It should be equipped with non-interchangeable connections. Care should be taken to ensure that all inlets to air supplies for breathing are situated in uncontaminated areas and that continuity of supply is assured.

Hotplates, ovens and furnaces

Hotplates, ovens and furnances should not be installed adjacent to areas where flammable liquids are to be handled. Heat-resisting surfaces should be provided for free-standing equipment and appropriate insulation and ventilation arranged for any built-in items.

Refrigerators

Only those refrigerators designed or modified for the storage of flammable liquids are suitable for this purpose. The storage of volatile, flammable liquids in ordinary refrigerators has led to serious explosions.

VENTILATION

General

Good general ventilation is essential for all laboratories and chemical stores. The design should take account of all factors such as fume cupboards, opening of windows, energy losses, etc., which can have a marked effect on ventilation characteristics. It is normally desirable to maintain laboratory air pressure below that of access areas. The specification of the system provided will depend on the nature and scale of the work but for most laboratories it should provide between three and ten air changes per hour. The laboratory heating system should be adequate to meet these requirements.

Fume Cupboards and other containment systems

The design of fume cupboards is complex. Readers are referred to the Society's guidance on laboratory fume cupboards for a more detailed treatment of the subject.

The design of any containment system should take account of the nature of the intended operations. The prime requirement of all such systems is that the air people breathe should be safe and without risk to health.

It should be remembered that the containment efficiency of a fume cupboard depends not only on its design but also on its siting. Operational aspects such as the amount, type, disposition and use of equipment in the cupboard will have a significant effect on performance.

Fume cupboards should be constructed from materials suitable for their intended application. Nevertheless the nature of the work carried out within most fume cupboards changes during their lifetime and it is sensible to provide at the outset for the most aggressive environment likely to be encountered.

In the design of fume cupboards it may be desirable to consider some form of energy conservation. The design should ensure that the system remains safe in the event of failure.

There are certain very toxic substances (eg substances which are very biologically active) for which total containment systems, such as glove boxes are required.

Operational aspects of containment systems are covered in Section E. Systems which rely on filtration and re-circulation should be considered with particular care.

This Section should be considered in conjunction with Section D.

PERSONAL PROTECTION

It is prudent to require the wearing of eye protection at all times. Safety spectacles conforming to BS 2092 or offering an equivalent degree of protection should be provided and used. Indeed they must be provided if:

a) the work carried out falls within one of the specified categories defined in the Protection of Eyes Regulations 1974

b) the wearer has monocular vision

c) the optical prescription is such that the centre of the lens is too thin for safety.

Wearers of contact lenses should be advised that these can interfere with effective eye irrigation in the event of chemical splashes, although there is evidence that they afford some protection against mechanical damage to the eye. Nonetheless additional eye protection should be worn. It is advisable that in the event of chemical splashes first aid should be limited to irrigation with clean water. The removal of contact lenses should be undertaken only by qualified medical staff.

Full eye protection (ie goggles) must be worn where there is a risk of splashing (eg when pouring corrosive materials from large containers). Face shields offer frontal and side protection from splashing and flying fragments but consideration must be given to whether they can safely be used alone or should be worn over goggles.

Every laboratory should have an adequate supply of gloves appropriate to the hazards present (eg heat, cold, corrosive materials and mechanical abrasion). If gloves are provided to protect against cold (eg when handling liquefied gases) they should be loose fitting so that they can be flicked off if cold liquid gets into the glove. The choice of chemical-resistant gloves depends on the materials to be handled. Useful reference charts can be obtained from manufacturers. Contaminated gloves must be washed before removal and not allowed to contaminate skin or surfaces such as door handles, switches, taps, handkerchiefs, pens or notebooks.

Staff should wear laboratory coats and these should be regularly laundered. The design of coats should be such that they can be quickly removed. Rubber or PVC protective aprons should also be worn when large quantities of corrosive or dangerous chemicals are being handled. Operations involving the handling of large amounts of hazardous substances may necessitate the use of impervious suits. In such circumstances it may also be necessary to use Wellington boots, and to wear outer garments over them to prevent ingress of spilled liquids. Where necessary appropriate safety footwear should be worn at all times.

Respiratory protective equipment may be required where effective ventilation or containment cannot be guaranteed. The choice may range from simple dust masks to self-contained breathing apparatus and will depend on the nature of the risk.

All forms of respiratory protection should be approved, used and maintained to the relevant British Standard and marked accordingly. A personal issue should be provided.

Areas of the laboratory where the use of particular personal protection is required must be identified and clearly indicated.

OPERATIONS UNDER REDUCED PRESSURE

When vacuum equipment is in use it should be adequately screened from all personnel. If flammable or toxic materials are involved the work should be carried out in a fume cupboard. It is essential that all glass components of vacuum apparatus should be of good quality and free from cracks or flaws. Thin-walled and/or non-spherical flasks must not be used as receivers in vacuum distillation units.

Laboratory filtrations are commonly carried out with porcelain or sintered glass filters supported on Buchner flasks which are evacuated by means of bench water-jet vacuum pumps. Such flasks can implode and they should be enclosed (eg in fine metal mesh cages). This also applies to glass desiccators for vacuum drying solids. All vacuum lines should be protected against the accumulation of liquid by the provision of adequate traps.

USE OF CENTRIFUGES

The attention of staff should be drawn to published codes of practice on centrifuges (see Section F).

Bench centrifuges should be firmly fixed and adequately shielded so as to prevent the ejection of broken tubes etc. Centrifuges should be sited where the vibration they produce will not cause bottles to fall from shelves. Centrifuges, like all electrical equipment, should be correctly earthed. Due account should be taken of the hazards of centrifuging flammable materials. For example, centrifugation of such material may produce vapours/mists above the LEL. In cases where it is not possible to run a centrifuge free from explosive vapour mixtures it is desirable to incorporate a system of purging and diluting with an inert gas such as nitrogen. All centrifuges should be equipped with interlocks to prevent the cover being opened when the rotor is in motion. However, many older machines are not so equipped. Where such interlocks are not fitted a warning notice stipulating a minimum delay period between switching off and opening the cover should be clearly visible.

The components of high-speed centrifuges must be regularly inspected in detail and any necessary maintenance work must be performed before use. Rotors and similar parts of the apparatus are subject to considerable stresses and may not give warning of incipient failure.

Before it is set in motion the balance of a centrifuge should be ensured by equal distribution of the load.

Special consideration should be given to the possibility of generating aerosols especially when centrifuging biological materials. In such cases special control measures will be required.

WORKING ALONE

Lone working may place the laboratory worker in a position of danger and should be discouraged. Where it is the policy of an establishment to allow lone working under certain prescribed circumstances it should be authorised only by a senior manager who should give permission in writing, naming the person involved, the place, the time and the nature of the work to be done. The person to whom permission is granted should be fully experienced in the work and should be forbidden to undertake any tasks which might lead to situations where a second person would be needed to render assistance.

Consideration should be given to arranging visits to, or other checks on, any persons required to work alone. However, such arrangements should be regarded only as back-up and in no way afford adequate protection by themselves.

UNATTENDED OPERATIONS

Experiments left running unattended (for example at night or over the weekend) can present hazards, particularly when flammable and toxic substances are involved. In addition to risks to personnel there is the possibility of damage to buildings and equipment, eg by flooding arising from faulty water lines. Ideally control devices should be provided, preferably with automatic trips to render equipment safe in the event of breakdown. It may be advisable to install relay trips to ensure that equipment remains shut down after a temporary electrical supply failure. Such devices should be tested thoroughly and regularly to ensure that they are reliable.

Where it is necessary to rely upon monitoring by a person patrolling a laboratory he or she must be adequately trained and should have clear written instructions on how to shut down equipment in case of accident or malfunction. The person patrolling should be able to make urgent contact with the person in charge of the operation or a named deputy. If the experiment requires the use of a fume cupboard there should be a means of alerting the external personnel should the extraction system fail.

THE DISABLED PERSON

The Disabled Person (Employment) Act 1944 and the Employment of Disabled Persons (Amendment) Act 1956 require employers of more than twenty persons to employ 3% who are Registered Disabled Persons.

Employers and employees have a duty under the HSW Act to protect the health and safety of employees and other people who may be affected by the organization's activities. Clearly in view of this statutory obligation it will be necessary to examine closely the work activities of employees known to be disabled. Common law requires an additional degree of care for the protection of any employee with known disability. This may, of course, necessitate an alteration in the type of work carried out by an employee who develops a disability or whose known disability increases. Thus for example, it may eventually be necessary to transfer employees with progressive disabilities from laboratory jobs to more suitable posts when their disabilities have become so far advanced that the employees become a hazard both to themselves and to others despite the provision of appropriate aids.

Each disabled employee should be considered as an individual case. As far as is reasonable the employer should take a positive approach towards providing an environment in which the disabled person can perform his or her work safely.

Managers should be assisted and encouraged to learn about the various aids which would enable disabled employees to work safely and effectively. There are several specialist organisations that provide advice on these matters. However, it should be noted that many of the technical aids that are available have been directed to applications within offices. There is a need for initiative to be shown in developing such aids for use in laboratories.

Particular attention should be given to emergency procedures and their suitability for disabled employees.

MAINTENANCE AND INSPECTION

The importance of maintaining laboratory equipment, services and control procedures in good order cannot be over-emphasised. Before a new item of equipment is acquired suitable procedures for its maintenance should be determined. Problems that may be produced by its installation should be considered, eg overloading of power supplies. These matters should be investigated in collaboration with appropriate staff.

It is important to remember that insurers may need to be notified of changes to laboratory equipment (eg pressure vessels).

Regular inspections should be undertaken to ensure that laboratory equipment, including protective equipment, remains safe and adequate for the work being performed. All electrical equipment must be regularly inspected to ensure that it is adequately insulated and earthed and has no obvious defects such as loose cabling. It is also important to check that ventilation systems remain functional and adequate. Inspections should take place at regular intervals and should, for example, seek to ensure that standards of tidiness and cleanliness are satisfactory, that corrosive materials are properly dispensed and that only minimum stocks of combustible materials are held.

Safety procedures should be regularly scrutinised by senior managers and, where appropriate, Safety Representatives, to ensure that procedures remain not only viable but as efficient as possible.

UTILITIES

(See also Section D)

Electricity

The main hazards arising from the improper use of electricity are shock, burns, the ignition of flammable materials by short circuit or switchgear spark, and the risk of damage to instruments by faulty wiring or operation.

Installation and maintainance of electrical services are usually the responsibility of a service department employing a qualified electrician. Nevertheless, the laboratory manager should ensure that a competent person checks all electrical equipment (including earthing systems, switches, sockets, plugs etc.) at regular intervals and carries out repairs or replacements when necessary. Residual current devices such as earth leakage trip units are recommended in addition to other precautions. All temporary circuits should be installed by competent persons. In especially hazardous situations the use of reduced voltage (eg 110v) supplies centre-tapped to earth may be preferred.

The laboratory manager must ensure that staff are trained in the operation of the electrical system installed. Staff should never attempt to modify or repair any electrical equipment or wiring but should report any faults to their supervisor so that they may receive prompt and skilled attention. Access to high voltage facilities (such as 11kv switch gear) is restricted by law.

It is important that an appropriate procedure be provided to switch off the supply to the laboratory in the event of fire since the fire-fighting operation might create fresh hazards by causing short circuits. All employees should know where isolation switches are located.

Fuel Gases

It should, be noted that colour coding is not internationally consistent. Therefore special attention should be paid to imported equipment.

Almost all of the gas now employed in laboratories as burner fuel is natural gas (methane) although propane and butane from cylinders may also be encountered. Acetylene and hydrogen are also used (eg in atomic absorption equipment where speciai requirements apply). The use of oxygen instead of air as an oxidant (eg in glassblowing) requires enhanced ventilation because of the formation of nitrogen oxides.

All fuel gas lines (and other services) should be colour-coded for identification and labelled at intervals.

As with electricity supplies all maintenance and repairs of service pipelines and equipment should be undertaken only by trained maintenance staff. The system should be inspected at regular intervals with particular attention being paid to any flexible tubing (for example, from bench cocks to Bunsen or other burners).

Everyone should be aware of the need to shut off all equipment including burners and pilot lights should the gas supply be interrupted for any reason.

Water

All pipes, valves and connections in water cooling systems should be inspected at intervals and prompt action must be taken to remedy faults. Flexible tubing should be adequately secured and drain connections should be well anchored to avoid flooding should water pressure vary. It should be noted that water pressure can rise dramatically at the end of the working day.(Water should, of course, be supplied to the bottom end of condensers).

Consideration should always be given to hazards arising from coolant failure.

Drainage

The laboratory manager should prepare rules about discharge of waste into sinks including authorised routes for disposal and should ensure that these are understood and followed by all employees.

Maintenance of drainage systems is especially important in chemical laboratories as construction materials can be affected by corrosion and solvent attack. Maintenance staff must not be put at risk from such chemicals or their effects. Where drainage traps, filters etc are fitted a routine for removal and disposal of collected waste must be put into operation.

Fume cupboards

Readers are also referred to the Society's guidance on laboratory fume cupboards.

Fume cupboards should not be regarded as convenient disposal routes for toxic or flammable waste gases and vapours. Whenever possible recourse should be made to removing noxious effluent as part of the experiment, eg by scrubbing, chemical absorption etc. Unnecessary chemicals should not be kept in fume cupboards that are used for experiments. The risk of fire caused by the use of portable electrical equipment such as hotplates, stirrers and heating mantles, none of which is likely to be flameproof, should be taken into account.

No fume cupboard achieves complete containment and even a well designed system will deteriorate with time or if equipment installed in it impairs its aerodynamic characteristics.

Fume cupboards should not be used where total containment is required.

The laboratory manager should ensure that users are aware of the limitations of fume cupboards and that all containment systems are maintained in satisfactory conditions.

PURCHASING PROCEDURES, AUTHORISATION AND ACCESS TO HAZARDOUS SUBSTANCES

The receipt and supply of hazardous substances should be the responsibility of trained and experienced stores personnel. A system of management control involving supervision by qualified and experienced staff must be provided to ensure that access to such substances is both safe and legal. For example, radioactive substances should be ordered only with the authorisation of the appropriate Radiation Protection Adviser and in accordance with the licencing requirements for the premises. These materials should remain under the control of an appropriately designated or authorised person acting within the local Code of Practice. Other highly toxic substances should be ordered only with the consent of authorised and skilled personnel and should be issued only to approved persons. Due regard must be paid to legal requirements (Section F).

Substances on the controlled list of the Misuse of Drugs Act must be stored in accordance with legal requirements. These materials must be under the control of personnel holding specific Home Office authorisation unless they fall within the general definitions with the Act. For most chemical laboratories such specific authorisation will probably be required.

Similar though less stringent requirements apply to the storage and issue of scheduled poisons.

Procedures for the issue of hazardous substances should ensure that the persons to whom they are supplied are aware of the hazards. They should also be fully conversant with the first aid requirements which may be specific to that material. For example, cyanides or hydrofluoric acid should be issued only to persons who are familiar with the necessary handling precautions and who have the appropriate antidotes.

Transactions involving hazardous substances must be fully recorded. Such records should show authorisation requirements and the scale and frequency of use. They should enable quick identification of the amounts and locations of such materials at short notice. This is of particular importance in respect of substances controlled by law including radioactive materials, poisons, drugs and certain carcinogens.

WASTE DISPOSAL

Disposal of laboratory waste can be achieved by techniques involving dilution, neutralisation, landfill or incineration, or by treatment by a specialist service. It should be remembered that accumulation of toxic material in the domestic sewage system, to which many laboratory drains are connected, can be hazardous to maintenance personnel. Furthermore the proper biological balance in the sewage treatment works can be seriously affected by changes in pH resulting from acid or alkaline effluent. The presence of heavy metallic and other toxic elements is also detrimental. Strict limits on the concentrations of these substances must be observed (see Section F). Disposal of acids and alkalis can be achieved by extensive dilution with water or, if the quantities are substantial, by prior neutralisation.

The Control of Pollution (Special Waste) Regulations 1980 make it an offence to deposit waste on unlicensed land if the waste is poisonous, noxious or polluting and is liable to give rise to an environmental hazard. Organizations that possess a sizeable area of waste land adjacent to the laboratory may be able to burn off flammable liquids on shallow metal trays. However the Local Authority must be consulted. Only small volumes should be burnt at a time and the person carrying out the operation should be accompanied by a colleague equipped with a fire extinguisher. Both should wear goggles and proctective clothing. N.B. regulations require the licensing of certain incinerators.

The risk of creating a nuisance must be borne in mind and, accordingly, volatile liquids with unpleasant smells should not be allowed to evaporate in the open.

Suitable covered containers should be provided in the laboratory or close to it for the collection and segregation of noxious and flammable waste. Care should be taken not to produce in these containers reactive mixtures that will in themselves create fresh hazards (chlorinated solvents, for example, should be kept separate from ketones). Legislation requires that the amounts of substances disposed of and the methods employed are officially recorded.

The disposal of radioactive waste is controlled by licence from the Radiochemical Inspectorate of the Department of the Environment.

RECORDING INFORMATION

The COSHH regulations require that certain records are kept. Readers are referred to the Society's detailed guidance on COSHH in laboratories.

An essential part of all experimental work is the recording of the methods used and the results obtained. The accuracy of such records is essential for the success of the experimental work; it may also be invaluable in the maintenance of safe practice or in the investigation following an accident. Staff should be required to maintain accurate laboratory records on a day-to-day basis. Both these records and the reports subsequently based upon them should preferably be kept in duplicate. If possible copies should be kept in different locations to minimise the risk of loss by fire or other mishap. One approach is to keep a microfilm copy in another location.

LEGISLATION AFFECTING CHEMICAL LABORATORIES

1. Health and Safety at Work Etc. Act 1974 (HSW Act)

The HSW Act contains provisions for securing the health, safety and welfare of persons at work, for protecting others and for controlling the keeping and use of dangerous substances. Furthermore it contains provisions for the control of certain emissions into the atmosphere and it defines general duties for the self-employed, persons in control of premises, and designers, manufacturers, importers and suppliers of articles or substances for use at work.

All employers must ensure so far as is reasonably practicable the health, safety and welfare at work of their employees. In particular employers should also:

a) Provide and maintain plant and systems of work that are safe and without risk to health

b) Make arrangements to ensure that the use, handling, storage and transport of articles and substances are safe and without risk to health

c) Provide such information, instruction, training and supervision as is necessary to ensure the health and safety of his or her employees

d) Provide and maintain a safe place of employment and safe means of access and egress that are without risk to his or her employees

e) Provide and maintain a safe working environment without risk to his or her employees with adequate facilities and arrangements for their welfare.

Employers with five or more employees must prepare a written statement of their general policy, organization and arrangements for health and safety at work. They must keep it up to date by revision and bring it to the attention of their employees.

Employers and the self-employed must ensure so far as is reasonably practicable the safety of persons not in their employment (for example, students, members of the public, visiting contractors) who may be affected by their activities at work.

Persons in control of premises also have duties towards others who are not their employees but who use the premises as a place of work. These duties require the person in control to ensure that the premises (including its entrance and exits) and plant or substances provided for use are safe and without risks to health so far as is reasonably practicable.

Employees must take reasonable care both for their own health and safety and for that of other persons who may be affected by their acts or omissions at work. They must co-operate with their employer so far as is necessary to enable him or her to comply with the requirements of the Act.

Designers, manufacturers, importers and suppliers of articles or substances for use at work must ensure that, so far as is reasonably practicable, they are safe when used. They must test articles for safety in use or arrange for such testing to be carried out. They must also supply information about the safe use of the article.

Anyone who installs or erects any article for use at work must ensure that, so far as is reasonably practicable, it does not constitute a risk to health and is safe for use.

The Act and other relevant statutory provisions are administered and enforced by the HSE and Local Authorities.

2. Regulations made under the HSW Act

2.1 Safety Representatives and Safety Committees Regulations 1977

These provide for recognised negotiating Trade Unions to appoint Safety Representatives and for the establishment of safety committees.

2.2 the Health and Safety (Genetic Manipulation) Regulations 1978

The Regulations provide that persons should not carry out genetic manipulations unless they have previously notified the HSE and the Genetic Manipulation Advisory Group. The Regulations extend the meaning of "work" in Part I of the HSW Act to include any activity involving genetic manipulation and any person so involved.

2.3 The Safety Signs Regulations 1980

The Regulations provide that certain safety signs should comply with British Standard BS 5378 Part 1, 1980, and that certain safety signs are not to be used at places of work except to provide specified health or safety information or instruction. The Regulations also provide that signs used for regulating traffic at places of work shall be the appropriate signs under the Road Traffic Regulations Act 1967.

2.4 The Health and Safety (First Aid) Regulations 1981

The Regulations place a general duty on employers to make adequate first-aid provision for their employees if they are injured or become ill at work.

2.5 The Health and Safety (Dangerous Pathogens) Regulations 1981

The Regulations relate to certain dangerous pathogens listed in a Schedule to the Regulations.

The Regulations prohibit the keeping, handling and transportation of a listed pathogen unless notice is given to the HSE at least 30 days in advance. The carrying on of a diagnostic service likely to involve a listed pathogen is also prohibited unless similar notice is given. The details to be included in a notice are set out in Schedules to the Regulations. The Regulations require the HSE to pass the information it receives to the appropriate Health Minister.

In addition, for the purposes of Part I of the HSW Act, the Regulations extend the meaning of "work" to include any activity involving the keeping or handling of a listed pathogen. In relation to such activity they provide that Section 3(2) of the HSW Act (which deals with the general duties of employers and self-employed persons to other persons who are not their employees) should be applied as if the reference to a self-employed person included a person who is neither an employer nor an employee (for example a student).

2.6 The Classification, Packaging and Labelling of Dangerous Substances Regulations 1984

The Regulations impose requirements in respect of the containers in which prescribed dangerous substances are supplied, the information to be shown on containers and the method by which the containers are to be marked or labelled.

2.7 Reporting of Injuries, Diseases and Dangerous Occurrences Regulations 1985

The Regulations provide for the reporting of accidents resulting in death or major injury to persons arising out of or in connection with work. Certain defined dangerous occurrences are required to be reported in a similar way. They also require records to be kept of accidents and dangerous occurrences, and of certain claims made in relation to industrial diseases.

2.8 Ionising Radiations Regulations 1985

2.9 Control of Asbestos at Work Regulations 1988

2.10 Asbestos (Prohibitions) Regulations 1988

2.11 Control of Substances Hazardous to Health ("COSHH") Regulations 1988

The Regulations relate to work involving substances which are defined as being hazardous to health. The Regulations require that employers do not carry out any work involving substances hazardous to health unless a suitable and sufficient assessment has been made of the risks to health created by the work and the measures necessary to control exposure to substances hazardous to health. The employer must ensure that such exposure is prevented or controlled. Control measures must be maintained, examined, tested and properly used. In certain circumstances monitoring of exposure to substances hazardous to health and health suveillance should be carried out. Employers must provide their employees with such information, instruction and training to ensure that they know the risks to health created by exposure to substances hazardous to health and the precautions which must be taken.

2.12 The Electricity at Work Regulations 1989 (from 1 April 1990)

3. Other relevant legislation

3.1 Explosives Acts 1875 and 1923

These Acts contain provisions controlling the manufacture, storage, supply and transportation of explosive substances. Premises must be registered and a person certified by the Chief of Police as a suitable person to keep explosives. They are administered by the HSE, Police and Local Authorities.

3.2 Petroleum (Consolidation) Act 1928

The Act requires licensing of storage of petroleum and certain other flammable substances. It is administered by the Local Authorities.

3.3 Public Health Acts 1936 and 1961

These include requirements concerning waste and sewage disposal, water supplies, building regulations, and trade effluent. They are administered by Local Authorities.

3.4 Disabled Persons (Employment) Act 1944, and Employment of Disabled Persons Amendment Act 1956 (See Section E)

3.5 Radioactive Substances Act 1948 and Regulations

These concern requirements for transport of radioactive substances.

3.6 Therapeutic Substances Act 1956

The Act requires that a licence be held for the purchase of certain substances. It is administered by the Department of Health.

3.7 Clean Air Acts 1956 and 1968

These contain provisions for the control of air pollution. They are administered and enforced by local authorities.

3.8 Radioactive Substances Act 1960

The Act concerns requirements for the control or disposal of radioactive waste. It is administered by the Department of the Environment. Inspection is undertaken by the Radiochemical Inspectorate in England and Wales, and by HM Industrial Pollution Inspectorate in Scotland.

3.9 Factories Act 1961 and Regulations

These contain provisions for health and safety in factories. The requirements of the Act are being replaced progressively by regulations made under the HSW Act. Certain laboratories (eg process laboratories) may come within the definition of a factory. In laboratories where the Factories Act does not apply its requirements may still be used as a guide where appropriate. A list of relevant regulations is given below:

- Abrasive Wheels Regulations 1970

– Highly Flammable Liquids and Liquefied Petroleum Gases Regulations 1972
– Protection of Eyes Regulations 1974.

3.10 Offices, Shops and Railway Premises Act 1963

The Act concerns requirements for general safety in defined premises. It is administered by the HSE and Local Authorities.

3.11 Nuclear Installations Act 1965 and Regulations

These require the licensing of nuclear reactors, sub-critical assemblies and certain other related experiments with nuclear fuels. They are administered by the HSE.

3.12 Gas Act 1965, Gas Safety Regulations 1972, Oil and Gas Enterprise Act 1984

These contain provisions for the control and safety of gas supplies. They are administered by the Department of Energy.

3.13 Fire Precautions Act 1971

This Act contains provisions for general fire precautions and means of escape in designated premises. It is administered by the relevant Fire Authority, with a few specified exceptions.

3.14 Misuse of Drugs Act 1971

The Act concerns restrictions on the production, supply and possession of controlled drugs. It is administered by the Home Office.

3.15 Poisons Act 1972

The Act restricts the supply and possession of non-medical poisons listed in the Poisons List Order. It permits supply to users approved by the Home Secretary. The Act contains requirements for labelling, storage and transport of poisons.

3.16 Control of Pollution Act 1974

This is an enabling Act which make provisions with respect to waste disposal, water pollution, noise, atmospheric pollution and public health. It is administered by Local Authorities.

3.17 Control of Pollution (Special Waste) Regulations 1980

These Regulations, made under the Control of Pollution Act, provide for notification and disposal procedures for special waste. They replace the Deposit of Poisonous Waste Act 1972 which is now repealed.

3.18 Occupiers Liability Act 1984

3.19 The Misuse of Drugs Regulations 1985

3.20 Medicines Act 1988

The Act restricts the sale, supply and transport of medicines especially those defined as "prescription only". It is mainly administered by the Department of Health.

BIBLIOGRAPHY

SECTION B – ORGANIZATION FOR SAFETY

Laboratory Organization

1. "Effective Policies for Health and Safety: a review drawn from the work and experience of the Accident Prevention Advisory Unit of HM Factory Inspectorate 1980", HMSO.

2. "Guidance Note on Employers' Policy Statement for Health and Safety at Work", Health and Safety Commission leaflet HSC 6.

3. "Guide to the HSW Act", HSE Health and Safety Series Booklet HS(R)6, HMSO.

4. "Safety Representatives and Safety Committees (regulations and guidance notes)", HSC 1988, HMSO.

5. Control of Substances Hazardous to Health Regulations 1988
– Approved Code of Practice HMSO 1988 (ISBN 0 11 885468 2)
– "COSHH Assessments – a step-by-step guide to assessment and the skills needed for it" HMSO 1988 (ISBN 0 11 885470 4)

6. HSE Guidance Notes EH40 and EH42.

7. Royal Society of Chemistry "COSHH in Laboratories" RSC 1989 (and an associated "Professional Brief" summarising the guidance).

Accident and Incident Reporting

1. Reporting an Injury or a Dangerous Occurrence", HSE 11 (rev), Health and Safety Executive.

2. "Reporting a Case of Disease", HSE 17, Health and Safety Executive.

Medical Services and Epidemiology

"Guidelines for Occupational Health Services" HSE Health and Safety Series Booklet HS(G)20, HMSO, London.

"Health Surveillance by Routine Procedures" HSE Guidance Note MS 18.

"Pre-employment Health Screening" HSE Guidance Note MS 20.

SECTION C – HAZARDS

Chemical Hazards

1. "Registry of Toxic Effects of Chemical Substances" (RTECS), National Institute of

Occupational Safety and Health, US Department of Health, Education and Welfare.

2. Sax, N.I., "Dangerous Properties of Industrial Materials", Van Nostrand, New York.

3. Steere, N.V. (Editor) "Handbook of Laboratory Safety", Chemical Rubber Publishing Co, Cleveland, Ohio.

4. "Hazards in the Chemical Laboratory", The Royal Society of Chemistry, London.

5. Bretherick, L. "Handbook of Reactive Chemicals Hazards", Butterworths, London.

6. "Patty's Industrial Hygiene and Toxicology", John Wiley & Sons, Inc.

7. Cooper, P., "Poisoning by Drugs and Chemicals, Plants and Animals", Alchemist Publications, London. (Out of Print).

8. Deichman, W.B. and Gerarde, H.W., "Toxicology of Drugs and Chemicals", Academic Press, London.

9. American Conference of Governmental and Industrial Hygienists, Annual List of Threshold Limit Values of Chemical Substances in Workroom Air, and the associated documentation of the Threshold Limit Values.

10. "Martindale, the Extra Pharmacopoeia", the Pharmaceutical Press, London.

11. Richardson, M.L., "Toxic Hazard Assessment of Chemicals", the Royal Society of Chemistry, London.

12. "Occupational Exposure Limits 19--", HSE Guidance Note EH 40 (Updated annually).

13. "Monitoring Strategies for Toxic Substances" HSE Guidance Note EH 42.

PHYSICAL HAZARDS

Lasers

1. British Standard 803, "Radiation Safety of Laser Products and Systems", 1983, British Standards Institution, London.

2. "Safety in Universities. Notes of Guidance Part 2:1 – Lasers", 1978, Committee of Vice Chancellors and Principals of the Universities of the United Kingdom. (under review).

3. American National Standard for the Safe Use of Lasers, ANSI 1981.

Cryogenics

4. British Cryogenics Council, "Cryogenics, Safety Manual – A Guide to Good

Practice", 2nd revised Edition, 1982, Mechanical Engineering Publications (London) on behalf of the British Cryogenics Council.

Pressure Reactions

5. "High Pressure Safety Code", 1975, The High Pressure Technology Association, London.
(N.B. New Regulations for pressurised systems are in the course of preparation).

Centrifugation

6. British Standard 4402, "Specification for Safety Requirement for Laboratory Centrifuges", 1982, British Standards Institution, London.

7. "User Guide for the Safe Operation of Centrifuges with Particular Reference to Hazardous Atmospheres", Institution of Chemical Engineers, Rugby. (Out of print).

SECTION D – DESIGN

Electricity

1. "Regulations for Electrical Installations", Institution of Electrical Engineers, London.

Fire Prevention

2. British Standard 5908, "Code of Practice for Fire Precautions in Chemical Plant", 1980, British Standards Institution, London.

3. "Industrial Use of Flammable Gas Detectors" HSE Guidance Note CS/1, HMSO, London.

Storage

4. "Topics in Safety", 1982, The Association for Science Education, Hatfield.

5. HSE Guidance Notes, HMSO, London. For Example:
CS2. "The storage of highly flammable liquids"
CS3. "The storage and use of sodium chlorate"
CS4. "The keeping of LPG in cylinders and similar containers"

6. "The storage of LPG at fixed installations", HS/G34.

Extraction and Ventilation

1. "Fume Cupboards in Schools", Design Note 29, Department of Education and Science, HMSO, London.

2. "Ventilation of Buildings: Fresh Air Requirement" HSE Guidance Note EH 22.

3. "An Introduction to Local Exhaust Ventilation" HSE HS/G 37.

4. Royal Society of Chemistry "Guidance on Laboratory Fume Cupboards" (in preparation).

Lighting

1. "Lighting at Work" HSE HS/G 38.

SECTION E – OPERATION

Eye Protection

1. British Standard 2092, "Specification for Industrial Eye Protectors", 1967, British Standards Institution, London.

Fire Fighting

2. "Gaseous Fire Extinguishing Systems: precaution, for toxic and asphyxiating hazards", HSE Guidance Note GS 16, HMSO, London.

First Aid

3. "First Aid at Work", HSE Health and Safety series booklet HS (R)11, HMSO, London.

SECTION F – GENERAL

1. Lees, R. and Smith, A.F., "Design, Construction and Refurbishment of Laboratories", 1984, Laboratory of the Government Chemist.

2. "Health and Safety in the Chemical Laboratory – where do we go from here?", 1984, Royal Society of Chemistry, London.

3. "Safety Audits – A Guide for the Chemical Industry" Chemical Industries Association.

4. "A Guide to Hazard and Operability Studies" Chemical Industries Association.

5. Kletz, T. "HAZOP and HAZAN – Notes on the Identification and Assessment of Hazards" Institution of Chemical Engineers.

6. Health and Safety Executive "Essentials of Health and Safety at Work" HMSO 1988 (ISBN 0 11 883977 2)

Appendix I

THE REPORTING OF INJURIES, DISEASES AND DANGEROUS OCCURRENCES REGULATIONS 1985

Statutory requirements are laid down in the Reporting of Injuries, Diseases and Dangerous Occurrences Regulations 1985, which came into force on 1 April 1986. Under these Regulations fatal accidents, major injury accidents and certain dangerous occurrences are to be reported directly to HSE. A dangerous occurrence is defined as one of seventeen prescribed situations. Of these the following may be particularly relevant to laboratory work:

i) the uncontrolled or accidental release or the escape of any substance or pathogen from any apparatus, equipment, pipework, pipeline, process plan, storage vessel, tank, in-works conveyance tanker, landfill site, or exploratory land drilling site, which having regard to the nature of the substance or pathogen and the extent and location of the release or escape, might have been liable to cause the death of, any of the injuries or conditions covered [by Regulation 3(2)] to, or other damage to the health of, any person.

ii) An explosion or fire occurring in any plant which resulted in the stoppage of that plant or suspension of normal work in that place for more than 24 hours, where such explosion or fire was due to the ignition of process materials, their by-products (including waste) or finished products.

iii) Electrical short circuit or overload attended by fire or explosion which resulted in the stoppage of the plant involved for more than 24 hours and which, taking into account the circumstances of the occurrence, might have been liable to cause the death of, or any of the injuries or conditions [by Regulation 3(2)1] to, any person.

iv) Explosion, collapse or bursting of any closed vessel, including a boiler or boiler tube, in which the internal pressure was above or below atmospheric pressure, which might have been liable to cause the death of, or any of the injuries or conditions covered [by Regulation 3(2)1] to, any person, or which resulted in the stoppage of the plant involved for more than 24 hours.

It is sometimes difficult to determine whether an occurrence is notifiable and when it should be reported. Whenever there is any doubt the local HSE office should be consulted.

Of the injuries and conditions covered by Regulation 3(2) the following may be particularly relevant to laboratory work:

i) the loss of sight of an eye, a penetrating injury to an eye, or a chemical or hot metal burn to an eye.

ii) Either injury (including burns) requiring immediate medical treatment, or loss of consciousness, resulting in either case from an electric shock from any electrical circuit or equipment, whether or not due to direct contact.

iii) Loss of consciousness resulting from lack of oxygen.

iv) Either acute illness requiring medical treatment where there is reason to believe that this resulted from exposure to a pathogen or infected material.

v) Acute illness requiring medical treatment where there is reason to believe that this resulted from exposure to a pathogen or infected material.

vi) Any other injury which results in the person injured being admitted immediately into hospital for more than 24 hours.

Notes

Notes